T0227428

Tele-Stress

Second Edition

By Stephen Coscia

CRC Press
Taylor & Francis Group
Boca Raton London New York

CRC Press is an imprint of the
Taylor & Francis Group, an **informa** business

CRC Press
Taylor & Francis Group
6000 Broken Sound Parkway NW, Suite 300
Boca Raton, FL 33487-2742

First issued in hardback 2017

© 2001 Stephen Coscia
CRC Press is an imprint of Taylor & Francis Group, an Informa business

No claim to original U.S. Government works

ISBN 13: 978-1-138-41223-1 (hbk)
ISBN 13: 978-1-57820-029-0 (pbk)

**Visit the Taylor & Francis Web site at
http://www.taylorandfrancis.com**

**and the CRC Press Web site at
http://www.crcpress.com**

To my mother, Mary Santullo Coscia, who taught me to work hard and serve others. I am still inspired by her life's example.

To my father, Anthony Coscia, the greatest man I have ever known. With him, I learned life's most meaningful lessons.

Acknowledgements

The Editors:

Special thanks to Tom Tracy, Tony Ferrara and Christine Kern. Their guidance and constructive criticism contributed to a coherent text. My wife, Veronica H. Coscia, provided the finishing touch to the final text and much love and support throughout the project.

The Artist:

Tom Tracy created the cartoon characters on pages 23, 25, 35, 58, 65, 76 and 77. His magical ability to draw my feelings and facial expressions is uncanny.

Special Thanks:

To Lauren Basham at SOCAP, Beverly Stoos, Regina Barr, Bernadette Baker, Robin Moriates, Ann Gallagher, Marilyn Kroger and Darlene Elkins, for their help with the call center stress survey.

Table Of Contents

Stop and Think

It's hectic! We're too busy getting things done to think. So we keep working, hoping to catch up. We never seem to get there. This condition adds self-inflicted stress to our lives. We're so busy reacting to crises and putting out fires that we don't permit ourselves to learn how to be more proactive.

If you are a call center professional, I bet your job is stressful and that you're up to your ears in work. Do you have a long term plan for yourself, or are you at the mercy of life's ebb and flow? If you need to do a reality check and refresh yourself with some new ideas and common sense principles, read on. The following story is a metaphor that mirrors many of our lives.

Two men were preparing to drive across a desert. They had just enough gas to make the trip in a straight run from where they were, to their destination. Less than halfway into their trip, they saw a tree. One of the men decided to climb the tree and get a different perspective on where they were going. From the top of the tree, he was alarmed to see a wide chasm in the land straight ahead. The chasm, which wasn't visible from ground level, looked a hundred miles wide and it would require re-routing their trip. Learning about this now was good fortune, because it was three-quarters into the distance of their trip. They wouldn't have had the gas to make it. Stumbling upon the chasm by surprise would have meant being stranded somewhere in the middle. Knowing that they did not have enough gas to drive the detour around the chasm, they decided to regroup and make a new plan.

Headed For Trouble

Let's stop for a minute so I can climb up this tree and get a different perspective on where we're going.

No time! We gotta keep moving.

Most call center professionals don't think they have enough time to stop, regroup and make a new plan. They have so much work to do that stopping, even for a minute or two, seems like a waste of their precious time. Nonsense! I urge you to start reserving a little time for daily thinking.

As you read this book, you'll learn that irrational thinking, or no thinking at all, is one of the leading contributors to stress. Making time to think rationally is a good start towards reducing the stress in your life and in your career as a call center professional.

If you're ready to begin, stop doing whatever it is you're doing, get comfortable, clear your mind of today's clutter and make time to think. Good luck!

I invite questions and comments about anything in this book. Feel free to contact me:

Stephen Coscia
P.O. Box 786
Havertown PA 19083
Phone: 610-658-4417
coscia@worldnet.att.net
www.coscia.com

Introduction

While writing my first book, *Customer Service Over The Phone*, I conducted a survey so I could learn what issues were most important to customer service professionals. This survey revealed three main areas of concern, in priority order: (1) selecting the correct words, (2) handling the stress of customer service and (3) solving complex problems. I focused on these three issues and, as a result, the first book's content has made it very popular.

About 18 months after my first book was published, I conducted a similar survey, asking the same questions to a much larger sample of customer service professionals. Again the same three important issues surfaced, but the distribution was different. Respondents to the second survey chose handling stress as their primary concern with selecting correct words and solving complex problems as second and third. This motivated me to conduct an exclusive call center stress survey. The survey results are in Chapter 5.

Tele-Stress includes some of my workshop text organized to help you understand what stress is, how it affects you both physically and emotionally and to provide some techniques for dealing with irate customers - the leading cause of stress.

To assist you in learning more about stress, the first chapter includes some short stories. These stories bring relevance and understanding to why we feel the way we do, when we get stressed.

This book continues with a chapter on how to deal with irate customers and the anger that results from these calls. Overcoming the impulse to become angry will take careful thought and application. Therefore I included a chapter on learning followed by a chapter on rational thinking. Also included is common-sense information on exercise, proper diet and positive self talk. The call center stress survey in chapter 5 reveals the primary causes of Tele-Stress and the most common post-stress behavior.

Customer behavior is changing. Based on my own experience, customers appear less patient and more demanding. In addition, technological advancements have made it more difficult for technically-challenged customers to keep up.

There is a divergence between what customers know and the pace at which things change. This divergence or gap keeps widening as the pace of change accelerates through time. This gap is frustrating to less-savvy customers, and their frustration is reflected in their attitude and in their tone of voice. CSRs must deal with this burden.

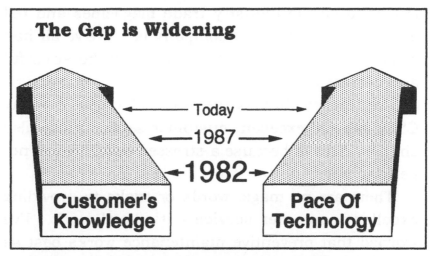

The Gap is Widening

Today

1987

1982

Customer's Knowledge

Pace Of Technology

During the last two decades, it's getting harder to keep up with the pace of technology.

As this divergence continues, an opposing phenomenon is occurring. The opposite of the divergence, between what customers know and the pace of change, is a convergence. The two issues that are converging are our customer's power and their access to consumer rights information. While the percentage of customers who exploit this phenomenon is relatively small, they certainly are vocal. In other words, they make our jobs very difficult.

These customers know what words and key phrases will push our buttons in an attempt to make us cave in to their demands.

Phrases such as, "If you don't give me what I want, I'll sue your company!" or "What's the name of your company's president? Maybe she can help me!" or "If you don't give me what I want, I'll mail letters to all the industry trade magazines and tell everyone how bad your company is!" We've all had calls from "customers from Hell." They're stressful and demoralizing.

I've learned that as the world changes, the way CSRs provide excellent customer service must also change. This is because a stressed out CSR will not perform well.

There are no magic words or tricks to providing excellent customer service - it's hard work. I've learned that preventive maintenance works best as a strategy for handling calls from difficult customers. As you get better at preventive maintenance, you'll experience less on-the-job stress because you'll prevent bad situations from getting worse.

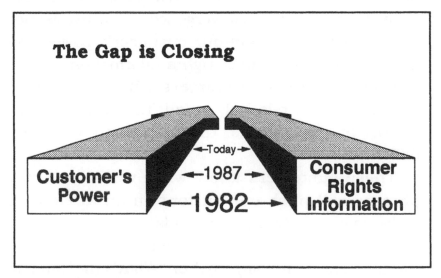

In recent years, some customers have exploited their easier access to consumer rights information.

Becoming skilled at preventive maintenance has a positive and cumulative affect on outcomes. Think of these proactive strategies as layer upon layer of bricks that, when neatly stacked, make up a formidable structure of better stress management.

Another trend that is affecting call centers includes some of the recent technological advancements. The purpose of new call center technology is to offer more choices of information fulfillment, improve customer satisfaction and expedite call resolution. Most new technologies achieve these objectives.

Many of today's call centers utilize interactive voice response (IVR) or some form of frequently asked question (FAQ) facility on an ACD. This new technology handles basic and routine customer inquiries with ease. These inquiries include: dealer referral, current account balance, shipping fulfillment date and tracking numbers, software version updates.

While traveling to call centers nationwide, I hear complaints from CSRs about the disproportionate number of difficult calls they handle. CSRs tell me that they used to handle a balanced mix of basic, routine and difficult calls. The basic and routine calls allowed for a much needed break between the difficult calls. Clinically, this break provided some necessary time for CSRs to recover subsequent to these adverse events.

Today, with IVR systems and FAQ facilities drawing the more routine and basic calls away from the CSRs in the call center, the only calls left for CSRs are the difficult ones. This condition leaves CSRs in a chronic and heightened state of awareness. In addition, handling repetitive difficult calls does not provide CSRs with enough time to recover between each call. This imbalance of workload makes working in a call center more stressful than ever.

Improved stress handling is a combination of knowing and applying the proactive things that work for you. What works for you may not work for someone else. We're all different. The key is to learn what works for you. This book will help.

Author's Note: I am not a doctor or a stress therapist. I am a customer service professional. What I have learned about stress in customer service is practical. The ideas and strategies in this book are not intended as a substitute for consulting with your physician. All matters regarding your health require medical supervision.

Chapter 1

Gaining New Knowledge

Writing a book about stress is a little precarious. This is because stress is such a subjective and relative phenomenon. No two people will respond to adverse situations in the same way. What one person considers a mild inconvenience is a major catastrophe for someone else.

Since people respond to adverse situations in different ways, the need for this book is relative based on who you are, where you've been, what you think about and why you think that way.

While reading this book, you might learn new concepts and ideas while others will only reinforce what they already knew. My objective is not to advise you, it's to get you to think. I want to share knowledge. Only by gaining new knowledge will we be able to adapt, reintegrate and re-make ourselves anew.

During my career in customer service, I have experienced much stress. This is because I have handled just about every customer service situation imaginable. Have I failed? You bet! Each mishandled customer interaction has been a learning experience for me. Along the way I have learned that stress contributed to some of my failures. So rather than allow stress to control me and the outcome of events, I decided to learn more about stress so I could learn to overcome it.

One of the most important things I've learned along the way is - *each stressful experience that I overcame better prepared me to handle the stressful experiences that lay ahead.* My threshold for stressful events kept getting higher and higher as long as I continued handling new events. The same thing will happen to you if you keep learning.

To gain new knowledge, I've read numerous books on stress, rational thought, and positive thinking principles. I have learned to handle and work along with stress, even strive on it, because I made the time to learn more about it.

Knowledge is the key to overcoming adversity. Fear and anxiety are often the result of ignorance about whatever is feared or makes you anxious. A simple analogy might be to imagine yourself in a dark room at midnight. Strange noises might scare you, until you turn on the light switch and reveal the source of the noise. Light fills the room with fact- finding, rational evidence that dispels fear and anxiety. Think of this chapter as your light switch.

Stress was designed to have a purpose in our lives. In this chapter, I'll explain how stress works, what instigates it and why it's normal and natural to experience stress.

What you don't know *can* hurt you. I believe that most CSRs don't know how or why their bodies react to stress. These stressful events cause some CSRs to hurt themselves, their families, their careers and their employers. Learning more about how our body reacts to stressful events is a positive first step towards overcoming the adverse affects of stress.

Knowledge will dispel myths and dogma. I manage to work along with stress because I know what it is, how it affects my body and why I shouldn't let it hurt me. I believe this same approach might work for you.

Stress And Real Danger

Larry, feeling rather tired and drained from a hectic day at the office, decided to take a stroll in the park across the street from his new apartment. He thought a short walk might do him good.

Suddenly, out of the corner of Larry's eye, he saw something move. A few feet away, a dog jumped out from behind a bush. Not knowing whether or not the dog was friendly, Larry started walking towards his apartment just in case. In an instant, the dog was snarling and growling and appeared to be charging straight for Larry. Now Larry was frightened, something was very wrong.

Fortunately, Larry is equipped with a mechanism that allows his body to immediately respond to dangerous situations like this.

Larry's cerebral cortex sends a message to his hypothalamus, which is where all the neural connections of the autonomic nervous system are gathered. The hypothalamus regulates all the involuntary organs such as the liver, heart, kidneys and so forth.

Larry's arteries constrict, his heart beats faster, blood is drawn away from his skin, his red and white blood cell count rises and his muscular tension increases.

Then, his liver is stimulated to release more sugar into the blood, which provides a burst of energy. His pancreas, after measuring the rise in blood sugar pours insulin into the blood enabling more sugar to enter cells in his body.

With the increase in heart rate, this optimized and energized blood is being carried throughout his body even faster.

Larry's digestive system shuts down to reserve all his body's resources for the crises at hand. His senses are heightened. The pupils in Larry's eyes dilate to let in more light, which helps him see better and avert obstacles in his path.

Homeostasis

Danger

Reaction

Recovery

As Larry runs up the steps to his apartment he can hear the dog's barks getting closer. Finally, as he slams the door behind him, Larry is safe. Now that the immediate threat has passed, his body begins a return to normal.

Stress And Perceived Danger

Bill is a customer service representative for the Windfall Computer Corp. He is one of Windfall's newest CSRs. Bill has had two weeks of training, working alongside a veteran CSR, and today is the first day that Bill is working on his own. His day started out OK. Mostly routine calls about technical issues that he has been trained to resolve. Then, he received a phone call from an irate customer.

Bill can hear the hostility and frustration in the customer's voice. Almost immediately, Bill got a sick nauseating feeling deep in his chest. As his metabolism increases, Bill's heart rate and breathing increase to facilitate his body's need for fast motion. Since he's unable to run, his hands and feet fidget about his cubicle. Bill is shifting in his chair. His eyes, now open wide, start darting around the room in desperation, as though there might be someplace to hide.

When Bill tries to interject with a word or two, he realizes his voice has become thin and raspy, making it almost impossible for him to sound confident, authoritative and efficient.

Homeostasis

Danger

Reaction

Recovery

Bill feels trapped in his cubicle, as he fights the urge to flee. He wants to throw something. He wants to run away. He just wants this situation to end.

The customer demands his money back, compensation for his lost time and a formal apology from the company president. Bill, not knowing what to do, transfers the call to a supervisor.

After transferring the call, Bill is an emotional wreck but there is no time to recover. His phone rings again. He begins his next call with a feeling of apprehension and uncertainty, which is clearly present to the caller.

Both Larry and Bill went through a four stage process. The first stage is *homeostasis*. Homeostasis is derived from two latin terms, *homeo* means "the same" and *stasis* means "lack of motion." Homeostasis simply means that the metabolism is at a normal rate.

The second stage is *danger*. At this stage, a hazardous or threatening situation is recognized and this condition instigates metabolic changes in the body. The third stage is *reaction*. This stage reveals a reaction to the threat. The final stage is *recovery* which occurs after the adverse event. Time for recovery is important because the body needs to return to a state of homeostasis. In call centers, there isn't always time to recover from a stressful event. This is a problem as some CSRs remain in a

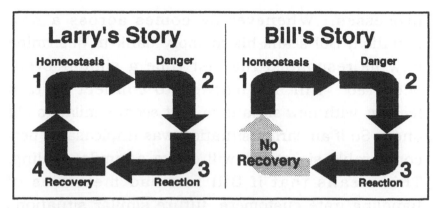

A usual four step stress cycle includes time for recovery. In call centers, the cycle might be reduced to three steps. Too often, CSRs must keep going without recovery time.

heightened state of stress due to repetitive difficult calls. This chronic condition is the best example of Tele-Stress.

Hopefully these two stories about Larry and Bill made you think about your own life's experiences. Surely at some point in your life you have been in a life threatening situation that caused you great alarm and concern. If you are still around to talk about that experience, your body's own survival mechanism came to your aid.

Bill's situation with the irate customer was not life threatening. Yet his body's metabolism increased to help him avert perceived danger rather than real danger. If Bill is going to change the way he reacts to irate callers, he must first change the way he perceives these situations. How can Bill do this? The answer lies within Bill's ability to change his beliefs, his thinking and finally his behavior.

Let's look at Bill's present belief and thinking

processes. Whenever he comes across a new situation, Bill scans his memory bank to determine how he responded the last time a similar event occurred. Bill has learned to connect certain feelings with new situations that seem similar to old ones. So if an earlier situation was unpleasant, new comparable situations will evoke a similar feeling. This means that if Bill has bad memories of handling irate customers, future similar situations will result in fear or anxiety that will invoke his stress mechanism. Bill's best remedy for breaking this negative pattern is to begin creating new experiences with positive outcomes.

These new positive experiences will eventually become the memories Bill will reference when he encounters subsequent situations. These positive outcomes will affect the way in which he handles future events as he realizes that he has overcome them in the past. This makes it possible for him to believe he can do it again.

Throughout life, people condition themselves to react to present-day events based on how these events were handled in the past. Start your new future by changing the way you perceive upcoming adverse events.

Chapter 1 Key Points

- Understand how and why your body's metabolism works. Identify your body's own stress signals.
- Our body's stress mechanism tends to react the same way whether we are threatened with real physical danger or emotional harm.
- Rational thinking is paramount in our challenge to handle stress.
- Start creating a memory bank of positive outcomes.
- Focus on "what to do" rather than "what is happening." Stay proactive.
- Changing the way you think starts with gaining new knowledge. Never stop learning.

Chapter 2

Irate Customers

A book about Tele-Stress would be incomplete without some information about the anger CSRs feel when handling calls from irate customers. Anger and irate customers are connected based on the survey results in chapter 5 of this book.

During a typical work day, you speak with scores of customers. Most are nice, a few are difficult, but you can bear it. The difficult ones don't push you over the edge, however each event is like a drop of water. It's annoying. A bother. Nothing you can't handle, just aggravation. One customer blames you for damage resulting from a rough freight carrier. Another customer is frustrated about fast-paced technology and hurls insults at you. Drip. Drip. Drip.

If you get enough of these innocent drops of water, eventually you'll have a bucket full of anger.

The next difficult customer just might be the one who tips the bucket over and - SPLASH! You're walking revenge!

There's no place for anger in customer service. Anger might seem appropriate as a way to keep the scores even. Yet, once the smoke clears and reality sets in, you'll realize that you were wrong. Then it's too late. You can't take back the words that were hurled like darts at your customer.

Anger only feeds revenge that feeds our anger that feeds revenge that feeds our anger - and so on. It's best to never get there in the first place, and containing your emotions is a good place to start. During each unpleasant experience, allow yourself a few seconds to think before you respond.

Take responsibility for your feelings. Other people don't make you angry. Other people can't make you do or feel anything. You choose to become angry because of what someone else does. You can't change other people's behavior, so you might as well change your own. It's best to stop blaming others for your circumstances and take personal responsibility for your feelings. If there are problems to resolve, fix the problems. Don't fix the blame.

The word "fix" is used in two different ways here. In the expression "fix the problems," the word fix means to repair. In the expression "fix the blame," the word fix means to set firmly in place. Fixing blame implies no responsibility on your part. You will not improve or benefit by seeing yourself as a victim who is at the mercy of life's ebb and flow. Practice optimism and positive expectancy. Overcoming anger includes learning to forgive. A grudge is too big a load to carry around.

Learn to suspend your angry reactions. Learn to use other positive techniques for venting anger and frustration (exercise, prayer, meditation, positive affirmations, talk with co-workers, rational thinking, positive self-talk). You will be rewarded for learning how to suspend anger. Life is so much more fulfilling when you remove anger from your usual thinking. On the same note, you will be punished for not learning how to suspend anger. A philosopher once said, "You won't be punished *for* your anger, you'll be punished *by* your anger."

In customer service, the fight or flight response manifests itself in interesting ways. While it's not possible to fight with or flee from our customers, an agents inability to cope might lead to psychological attempts to do so. Psychological flight might manifest itself as apathy or discourteous behavior. This type of behavior creates distance between an agent and his customer. This is contrary to the practice building closer relationships through empathy and genuine concern.

Psychological fight might emerge as aggressive or retaliatory behavior. This is unacceptable in an industry where it is essential to use restraint rather than retaliation.

When I recall how I consciously started thinking rationally about stress and its affects on me, a meaningful event stands out. One late afternoon I handled a call from a very difficult customer. It was the last call of my day. I left the office angry and frustrated. While driving home, I kept replaying this bad phone call over and over again, in my head. I blamed that customer for ruining my night. I was barely paying attention to my driving. I reached a sharp curve in the road and I lost control of my car. Fortunately, I didn't crash. My car did two 360 degree spins and then came to a screeching halt.

As I sat in the driver's seat, staring at the dashboard, I knew I was wrong for allowing myself to get so angry. I allowed myself to think irrationally. The expression "circumstances don't make a man, they reveal him" certainly rang true that evening. I was exposed as a man who didn't handle adversity well.

The events of that night have changed the way I think about stress. Even today, when I handle calls from irate customers, the memory of that night reminds me that I have a choice. These days, I choose to suspend my anger.

React or Respond

Thinking rationally about stress includes believing you have choices. In customer service, when customers get upset, you have two choices: you can *react* to what they have done or you can *respond*.

Anyone can react to an adverse event by simply

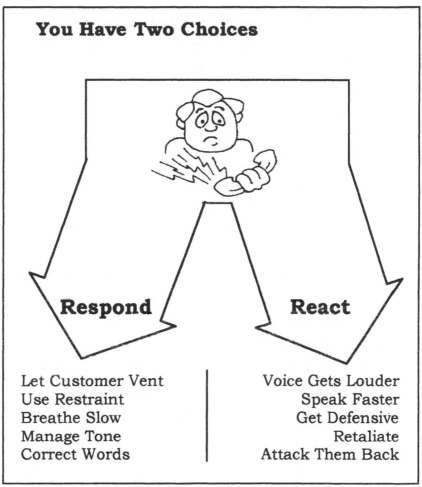

You Have Two Choices

Respond / **React**

Respond	React
Let Customer Vent	Voice Gets Louder
Use Restraint	Speak Faster
Breathe Slow	Get Defensive
Manage Tone	Retaliate
Correct Words	Attack Them Back

When an irate customer starts yelling, you have two choices. You can either react or respond.

doing what feels natural. If someone yells at you, you start yelling back. A reaction doesn't take much work or thinking on your part. In addition, a reaction is retaliation for a customer's bad behavior. You must learn that the customer's behavior is not the problem. The problem is whatever is causing the customer to behave that way. Fixing the real problem will inevitably fix the behavior.

If you choose to *react,* your voice will get louder, you'll speak faster, become defensive, and you'll retaliate. These reactions are inappropriate and unprofessional. The situation at hand will escalate out of control if you and the customer engage in a shouting match. In the end, you will lose. You won't be able to take back the angry words that you have already hurled at the customer. It's best to avoid these obstacles by not creating them in the first place.

A response to an adverse event takes a little more effort. If you choose to *respond,* you will have to think rationally about what is happening, consider your best options, and then execute a plan.

If you choose to respond, you'll let the customer vent, you'll breathe slow, you'll manage your voice and you'll use the correct words.

Let Customers Vent

Give customers time to vent. This is their side of the story. Even if their side of the story includes some bad behavior. Let them get it off of their chests. Customers feel better when they're given a chance to tell you their story.

When a customer vents, don't interrupt. If you do interrupt, he'll think you weren't listening and he'll go right back to the beginning of his story.

When an irate customer vents, he is climbing a hill, and when he gets to the top, he'll feel better. His story must be told in a contiguous sequence of events. When you interrupt an irate customer, he'll think you are not listening. Then he'll start over, except this time he might be even more irate because he believes he's speaking with someone who is not paying attention. Let him tell his whole story until he finally peaks.

Experienced CSRs are most prone to interrupt because of all the problems you've heard about and resolved. You get so good at identifying problems, that when a customer is halfway through a problem description, you already know what the solution is,

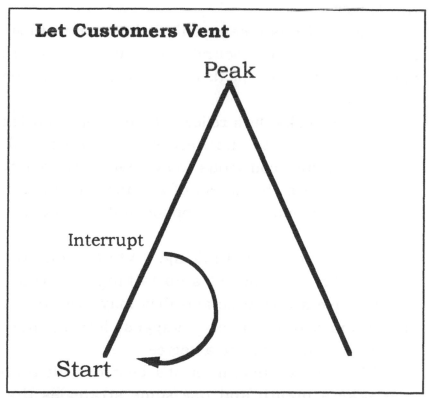

Let Customers Vent

Peak

Interrupt

Start

Venting is like climbing a hill. Customers don't feel better until they peak. If you interrupt, they'll start over.

so you cut to the chase and save the day.

But, it doesn't always work this way. Some customers will not appreciate your interruption. So what you thought was a helpful move, turns into a exasperating experience for the customer.

It's best to give the customer the extra 20 or 30 seconds to complete his thought and not interrupt, even though you know what the solution is.

An upset customer needs to vent to feel better. When people get stressed, it's usually because of a perceived danger or harm. The body reacts to avert the danger.

A stressed customer won't feel better until he knows he's safe and sound. So, when customers vent, getting to the peak is like getting to a safe place.

Once he peaks, he's relieved himself of a burden and placed it into the hands of a concerned, professional and courteous individual. This reality starts to relieve the stress of an upset customer, thereby giving you a greater opportunity for problem containment.

Some of you are thinking, "What about the person who just doesn't stop talking?" I know! Some people are verbose and they may need to say things more than once. I suggest letting these people have their say - to a degree.

With verbose customers, it's best to wait until they take a breath and use some strategies that might enable them to listen, rather than talk. This means, you must control the conversation.

These strategies include paying the customer a soft compliment or reminding them how precious their time is.

One of the best ways I know to control a conversation is to pay the other person a compliment. This will keep you talking and them listening. No one ever interrupts you when you are telling them how smart they are. These compliments must be subtle and sincere.

When customers vent, listen actively and take good notes. Then, using your notes, you can read the situation back to them. Good listening skills are important here. Good listening includes sounding like you're actively involved. People will share more with you when you're really listening. After a customer is finished venting, be sure to show empathy. This must be genuine. It's best to show empathy in such a way that lets the customer see who you really are.

Here's another subtle technique. Did you ever catch yourself saying "OK" as someone tells you a story. Saying "OK" is a benign way of acknowledging that you understand what's happening. This unintentional acknowledgement can be a problem for some customers. Don't say "OK" when customers vent. Some customers might turn this against you and say, "No, it's not OK!" It's best to say something like "I understand" or "I see."

Let's think for a minute about the difference between reacting and responding. You can undo all the good work you've done, if you react.

Human beings are reaction machines. Remember, when you're angry, you'll make the best speech that you'll ever regret. When things get hot, your goal is not to control their behavior - it's to control your own. We can choose to suspend our reactions to difficult situations. One way to do this is to acknowledge the customer's point of view.

Acknowledging the customer's point of view does not mean that you agree with him. It simply means that you accept it as one valid point of view among others. You're sending the message, "I can see how you view things."

Use phrases like, "You have a point there" or "I know what you mean" or "I understand what you're saying." Don't say these phrases mechanically. The spirit and intent with which you say them is just as important as the words.

Take a step to their side. Their side is the most valuable place for you to be. Seeing the situation through their eyes will help you better understand what's important to them and what obstacles stand in the way of agreement. Their side is also the safest place to be, because if you're on their side, it's difficult for them to attack you. It's also the most constructive place to be, because side by side you can work together to resolve differences.

Never give a customer the justification for his bad behavior, no matter how angry, difficult or abusive he gets. Here's why: When you behave the way he does, and react, and retaliate, you give him an excuse to continue being difficult. Always maintain your professional demeanor.

Don't stoop down to his level. One of our jobs is to absorb some of this verbal abuse, but only for the short term. Maintaining a courteous and professional demeanor contributes to improved stress handling, because it enables you to contain bad situations before they get worse.

Your Voice

When customers yell at you, you'll probably find yourself clearing your throat. You may need to clear your voice more than once and this is not only noisy, it's revealing. You're revealing that you are shaken-up and disturbed and this is not the message you want to send. Since stress will instigate a change in your voice, it's important to allow this change to occur in the background, without the customer hearing what's going on. If a customer starts yelling at you, try muting your phone for a few seconds, so the customer can't hear the changes in your breathing and in your voice.

The stress mechanism makes your voice sound thin and raspy. So, keep a glass of water or juice handy to lubricate your vocal chords. It's best to make sure your voice is working at an optimum level, especially if you're handling an irate customer.

Always speak with the voice of authority and confidence. Say your words with precise clarity and do not slur your speech.

Pay attention to the way you use vowels and consonants in your vernacular. Be aware of the soft

Slurred Phrases	English Translation
1) Hayadoin?	1) How are you doing?
2) Waitaminit!	2) Wait a minute!
3) Jeet?	3) Did you eat?

Slurring your speech makes you sound less intelligent, dull, poorly educated and lazy.

consonants such as H, Y, J and W, that cause you to slur from one vowel to the next vowel. Consonants should start and stop your vowels with deliberate precision. Doing this will enable you to send a positive and alert vocal image to a difficult caller who might be trying to rattle your cage.

Everyone's voice has some melody. Our vocal melody adds to our tone which conveys, excitement, inquisitiveness, surprise and statements of fact. One common flaw among telephone professionals is making their statements of fact sound like their asking a question. Our voice's melody for asking a question is different from the melody used for conveying a fact. Notice that when you ask a question, you voice typically resolves on a upward pitch. However, statements of fact typically resolve on a downward pitch. Making your statements of fact resolve on a upward pitch, as though you are asking a question, makes you sound uncertain or apprehensive.

I urge all telephone professionals to convey their facts in an appropriate manner and resolve these statements on a downward pitch. Maintain an authoritative tone that conveys your expertise and professionalism.

The Constructive Dialogue

This book contains many ideas and strategies that will help you to help yourself. It's best to start applying some of my ideas immediately. Find out which ones work for you and which ones need to be embellished to fit into your work environment.

One of the best ways to reduce stress in customer service is to avoid adverse experiences altogether. This means never getting there in the first place.

One proactive strategy that has worked for me is to engage an irate customer in a constructive dialogue. The purpose of the constructive dialogue is to divert the customer's attention away from the problem for a few moments. The content of the dialogue must be positive and it must be something that the customer is interested in or can relate to. When well-executed, a constructive dialogue will leave an irate customer with a more positive impression of you and your company. For example, here's a true story about one of my constructive dialogues.

One morning the receptionist at my facility alerted me about an angry customer who demanded to speak with a manager. It was 8:58 and I was getting ready for a 9:00 meeting, but I took the call. I allowed the customer to vent.

This customer was upset because, after numerous attempts during the previous day, he was unable to get through to our customer service department. Our lines were very busy at that time. I assured the customer that one of our CSRs would call him within the hour. Then I asked him for his phone number. He lived in area code 516, which I immediately recognized as Long Island, New York.

So, I asked him where he lived. He said, "New Hyde Park." I told him I was raised in Floral Park, which is a neighboring town. We exchanged a few niceties about high schools, local hang outs and closed the call amicably. That was a constructive dialogue.

A constructive dialogue might include the weather, current events or a geographical landmark specific to the customer's home city. I have also managed to include a few words about my son, which invited a customer to mention his children. When parents speak about their children, they do so with a sense of pride and satisfaction that comes through in their voice. The feeling of pride and satisfaction does wonders towards changing an irate customer's negative outlook.

The bottom line on helping yourself when handling irate customers is to stay calm enough to use a constructive dialogue. If you can remain calm, then you can consider what strategy might work best. Maintaining an attitude of grace under

Contain, Qualify & Correct

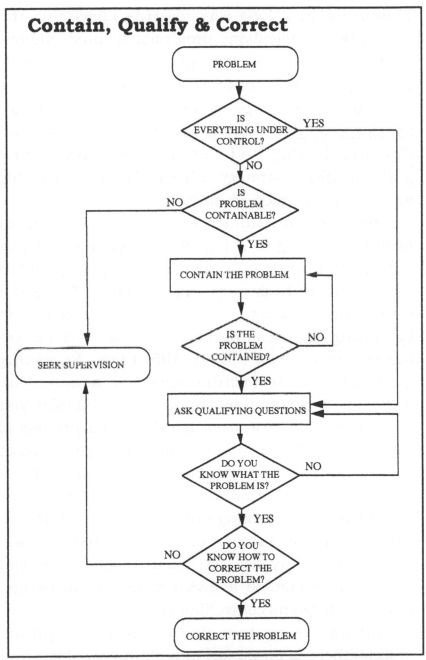

Contain means to keep within fixed limits. Problems can be contained using the strategies in this book. It's best to contain problems before you qualify them.

pressure goes a long way towards alleviating the daily stress a telephone professional must endure. When you get upset or get caught off guard, thinking is usually the first thing to go.

If you have a copy of my first book, <u>Customer Service Over the Phone</u>, review the chapters on *Contain, Qualify and Correct*. This systematic problem solving strategy will enable you to simplify and resolve complex problems.

The *Contain, Qualify and Correct* method of problem solving is a logical and common sense strategy that works. Contain means to keep within fixed limits or to prevent exacerbation. To keep a situation contained, use the strategies in the beginning of this chapter. When situations are under control, you may qualify them, by asking questions to determine what is wrong. The qualifying process of asking questions, enables you to learn what the problem isn't, using a process of elimination. After you have sifted through the most common causes, you can narrow down the possibilities. Asking questions allows you to control the conversation. Containing and qualifying a situation, will enable you to correct it. You must contain before you qualify and you must qualify before you correct. This method allows you to control events, rather than events controlling you.

During the contain process and the qualify process there is an option to seek supervision. You should reserve this option for the exceptions that may arise. Escalate legal threats or extremely difficult calls to your supervisor. Your supervisor has a broader scope of information to work from and can negotiate a mutually beneficial solution.

You can't get to the core of the problem until you defuse the customer's behavior. Irate customers want to fight. In this condition, customers may behave unpredictably. Using this method will yield more predictable results.

The *Contain, Qualify and Correct* flowchart on the opposite page illustrates the process and flow for improved problem containment.

Enigmatic Customers

Alan Chaney is a technical support rep for ABC Technology, a high-tech manufacturing company located in Pennsylvania. ABC manufactures computer peripherals. Everyone at the company is excited about a new model that was just released and ABC is offering an introductory low-cost upgrade option for customers who own older models. This way, customers who recently purchased an ABC product won't feel left out. The upgrade option was offered for a period of six months and the deadline was clearly posted on the company's website and trade magazine advertisements.

Three weeks after the expiration date, a customer called with an update request. When Alan informed the customer that the deadline had elapsed, the customer became upset. The customer griped that he had sent his upgrade form one month ago, but he had written "WA" on the envelope instead of "PA" which caused the post office to return his upgrade offer. Alan listened to the customer's gripe with empathy and genuine interest.

Alan thought it was odd that the post office return an envelope considering that the zip code is more pertinent than the state. Still, he kept an open mind and worked along with the customer.

Alan asked the customer if he still had the envelope and if the post mark date was visible. The customer answered "yes." "That's good news," said Alan as he suggested a win/win resolution. Alan suggested that he could make an exception, but ABC's financial controller would first need some proof of the mishap for final approval. Alan instructed the customer to simply fax a Xerox of the errant envelope. Then, Alan continued, he could attach the Xerox to the order form and everyone would be satisfied. The customer hesitated, then said that he would get back to Alan. The customer never did.

Alan was confused about why the customer didn't buy the upgrade. Alan wondered if the envelope ever existed or if it were just a story the customer had fabricated.

Have you ever handled a call that just didn't make sense? Calls such as there cause us to wonder whether or not a customer is telling the truth.

In the previous story, Alan did the conscientious thing by being fair, thereby ensuring a mutually beneficial outcome. Alan contained this event by not provoking the caller. In addition, Alan gave the customer every opportunity to achieve satisfaction.

I suggest that even though there was room for speculation on Alan's part, about the customer's honesty and intentions, it's best to assume all customers tell the truth. If CSRs make judgments about a customer's intentions or motivations, this will result in a breakdown in the good will that must exist between company and customer.

In telephone communication callers don't *take* meaning from your words, rather they *give* meaning to your words. Therefore, if a CSR prejudges, this will be subject to the caller's interpretation. If a caller believes he's being blamed and accused there is potential for an explosive event. I make it a policy to take all information at face value until facts prove otherwise.

It's very hard not to be judgmental when working with enigmatic customers. Denying the inclination to reveal a caller's motives takes rational thinking and mental discipline. While this sounds easy, it is not. Handling these situations is very hard work.

Since building good rapport is a key part of maintaining strong relationships, it's best to make no assumptions about a customer's behavior. CSRs should maintain the self esteem of others and practice the golden rule: treat others the way you would like to be treated.

How are enigmatic callers stressful? If a CSR jumps to conclusions or becomes judgmental, the result could be disastrous for the CSR and his employer. Calls like this might escalate to a legal matter. When it comes to stress, being proactive is the best strategy. Never get there in the first place.

The Dreaded Morning Call

On a sunny Thursday morning, Charles arrived at work in a great mood. He is a customer service representative at the Arrow Corporation. Charles has been having a good week so far. His upcoming weekend trip to the mountains is preoccupying his thoughts. He starts the day's work believing that it will be just like any other day.

He puts on his telephone headset and unconsciously logs onto the phone queue. He sees his incoming call LED light up, hears the click in his headset and greets his first caller in a perfunctory manner. The caller is angry and Charles winces as the customer yells.

Charles is taken by surprise and he jumps out of his lulled condition in a split second. He was not ready for this. He tries to plan a strategy, but it's too late because he can't think straight. Trying to recover from the verbal attack, Charles asks, "May I investigate the matter and call you back in a few minutes?"

hoping this will be enough time to regain his composure, recollect his thoughts and work more effectively. The customer says "No, I want an immediate resolution!" Charles is stuck. He can't think his way out of this situation. Charles struggles his way though the call and thinks to himself, "It's going to be a miserable day!"

In this story, Charles was not ready for the worst thing that could happen. He was operating as though he was on auto-pilot without much thinking or concern for what might happen next.

Early in my customer service career, I noticed a phenomenon that had a crippling affect on CSRs. This phenomenon is called the *dreaded morning call.*

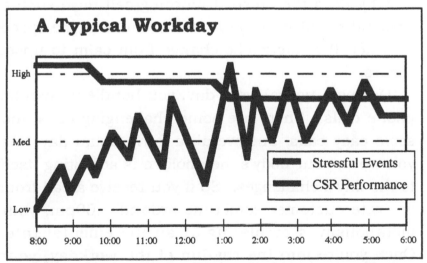

During a typical day, minor stressful events will not deteriorate your performance.

It's when one of the first calls of your day is a call from an irate customer.

These dreaded morning calls are stressful because of what happens inside our body the moment you pick up the telephone receiver and hear the irate customer's voice. In the morning, your body is in a state of homeostasis. Your metabolism is nice and steady. A call from an irate customer escalates your body's metabolic rate in a matter of seconds. Going from homeostasis to danger, first thing in the morning, affects your mental behavior in addition to your physical condition.

I have noticed that CSRs who experience a dreaded morning call, usually don't recover and return to their usual rate of productivity. These calls shake you up and they can ruin your day.

Why does the dreaded morning call have such a devastating affect on you? It's because of the speed in which the metabolic change from calm to upset occurs.

During a typical work day, you handle numerous phone calls. There are some challenging calls and some easy calls, but overall a fairly usual day. As you work, your body's metabolism is adjusting itself to the day's challenges. So if you receive a call from an irate customer sometime around 1:00pm, your metabolism is escalating at an incremental rate. Once you're into the rhythm of the workday, your metabolism has already been conditioned by the

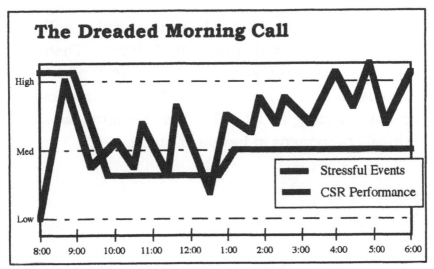

The Dreaded Morning Call

High

Med

Low

8:00   9:00   10:00   11:00   12:00   1:00   2:00   3:00   4:00   5:00   6:00

Stressful Events
CSR Performance

If the first call of your day is a dreaded morning call, your performance will deteriorate.

numerous calls you've already handled. So your heart rate and breathing have accelerated to an active pace. But, first thing in the morning, your metabolism is not conditioned.

So the fast escalation that you experience has a devastating affect on your behavior. You might find yourself thinking irrationally. Or you might find yourself not thinking at all and become overwhelmed.

I suggest that you get prepared for dreaded morning calls when you arrive at work in the morning. I advise a ten second adjustment, whereby you seriously ask yourself a question: "Am I ready for a dreaded morning call?" If the answer is no, then get serious and adjust your mental frame of mind.

Whether you get one or not, be ready. If you're

TELE-STRESS   53 ready and prepared for a dreaded morning call, you won't be caught with your guard down. Then, if it happens, you can start thinking rationally about what is happening instead of thinking irrationally, or not thinking at all. Rational thinking is a very critical component in handling stress well.

Chapter 2 Key Points

- •We always have a choice whether or not to become angry.
- •Take responsibility for your feelings.
- •Learn to respond rather than react when handling irate customers.
- •Let customers vent even though you hear their story numerous times daily.
- •When you get stressed, your voice becomes raspy. Try muting your phone, then clearing your throat so callers will not hear you.
- •Use the constructive dialogue to divert an irate customer's attention away from their problem for a few moments.
- •Always preserve the self esteem of callers, even when you suspect their intentions might not be honorable.
- •In the morning, be ready for the worst thing that can happen. Don't be caught off guard by the dreaded morning call.

Chapter 3

How We Learn

What does learning have to do with Tele-Stress? The answer to this question is; just about everything. Much stress stems from a CSR not having the correct skills to overcome and resolve adverse events. An unskilled CSR experiences much self-induced stress simply because he is incapable of reaching a desired outcome. This condition makes a CSR feel powerless, which makes his job even more stressful.

One of the biggest challenges in learning and gaining new knowledge is application. This means practicing what you've learned. The real value of my ideas and strategies won't be realized until you put them into action.

When you learn, you go through different stages of development. Abraham Maslow, the noted psychologist, identified four stages of learning. Each stage forces you to gain new knowledge and reap the benefits of your hard work. To make this meaningful, I will share a true story about my first customer service job.

One fine summer morning in 1982, I arrived at the office of my new employer. I was ready and eager to begin a career in customer service.

On that first day, my manager showed me around the entire facility. He showed me the warehouse, the shipping department, he showed me where the accounting department was, where the bathrooms were and then I was walked to a desk and the manager said: "Here's your desk and here's your phone. Just watch the other people working around you. Do what they do. Good luck."

Sound incredible? Actually that situation wasn't too uncommon back in 1982. Within sales and customer service departments, there was a belief that anyone could speak on the telephone. Just hire an individual, show him where his desk and his phone were, and let him talk to customers. Back then, customer service jobs weren't taken too seriously.

When I sat down and when my phone rang, I was *unconsciously incompetent* about how to do my job. I didn't know what my job was (unconscious) and I didn't know how to do it (incompetent).

When I sat at my desk and got settled, my phone rang. It was a customer looking for a back-ordered item. He wanted to know when the item would be shipped. I told him I didn't know. He didn't like my answer and started yelling about my company's poor customer service and that he was sick and tired of it. I asked the person next to me what I should do. He said that I ought to get the customer's name and phone number and call him back with an answer. After I captured the customer's information and put down the telephone receiver I had reached the next stage of learning. I was now *consciously incompetent* about how to do my job. Now, I had a pretty good idea of what my job was (conscious) and I also knew I still didn't know how to do it (incompetent).

During the following weeks, I started to learn everything I could about communication and people and customer service. I read books on communication, customer service, stress and positive thinking principles. There was much trial and error. I made mistakes and I started to learn from my mistakes.

Eventually, after much learning and application, I reached the next stage of learning and development. I became *consciously competent*. This means that I knew what my job was (conscious) and I was good at doing it (competent). I consciously selected the correct words as I worked with customers. I learned to slow down my rate of breathing and my rate of speech when customers yelled at me. I learned to focus on positive outcomes of events not on negative possibilities.

After months of applying these strategies over the telephone, it got to be a habit. And that's when I reached the final stage of learning. I became *uncon - sciously competent*. I barely had to think about what to do (unconscious) and still I performed very

Unconscious
Incompetence

Conscious
Incompetence

Conscious
Competence

Unconscious
Competence

well (competent). I automatically selected the correct words as I worked with customers. If a customer started yelling, I automatically started breathing and speaking slower. While my behavior was automatic, my communication style was always sincere and genuine. You see, sounding sincere and genuine also became automatic.

Those are the stages we go through when we learn. At the *first stage* of unconscious incompetence we're unknowledgeable and incapable of doing what needs to be done.

At the *second stage* of conscious incompetence we might feel a little anxious and apprehensive because we can see just how far we have to go. At this stage, it's important not to give up. This is where we have to dig deep, find our hidden resources and persevere. Positive thinking is critical at this stage. Visualize yourself doing the task. That's right. Create a mental picture of yourself doing whatever it is you're learning. At this stage, belief in yourself is important.

I've learned you must truly believe in a positive outcome before your behavior will make that positive outcome possible. Beliefs precede behavior. Let me tell a true story to prove my point.

Prior to 1954, breaking the four-minute mile was believed to be impossible. No runner had ever done it on record. Then on May 10, 1954, Roger Bannister broke the four-minute mile. Prior to that time, no other person believed it was possible.

However, between May 10th and the end of that year, 37 more people broke the four-minute mile. Did humans suddenly get faster? Probably not. What changed was their belief. They knew it was possible - so they did it. That's how important beliefs are in affecting our behavior.

At the *third stage* of conscious competence, we start to get better at doing our job. We start feeling more comfortable using the new techniques and strategies we've learned. Each stressful experience we overcome prepares us to handle the stressful experiences that lie ahead.

Then finally, at the *fourth stage* of unconscious competence, we find ourselves performing by habit rather than by conscious thought and application. We perform difficult tasks with grace and proficiency. It becomes automatic.

Inviting What Is Conveyed Towards Others

Portal Inc., one of the world's leading producers of high tech computers, had an internal problem with their forecasting software. This problem went unnoticed for weeks. Portal's buyers used this software to place orders with their respective suppliers. When the forecasting problem was first noticed, Portal 's buyers learned that the previous six weeks of purchase orders were incorrect. The forecasting software problem only gave the

EACH STRESSFUL EXPERIENCE WE OVERCOME PREPARES US TO HANDLE THE STRESSFUL EXPERIENCES THAT LIE AHEAD.

buyers visibility to about 60% of Portal's real production needs. This meant that Portal's buyers had to procure more materials - and fast.

When Portal's buyers phoned their suppliers to buy more materials, some suppliers were able to adjust their schedules and some were not. Whether a supplier could or couldn't adjust their schedule depended on production runs, retooling costs and raw material availability.

The suppliers who provided Portal with customized materials had the most difficult time adjusting their schedules due to the constraints of using special-order components. These special-order components had to be purchased ten weeks in advance. So, these suppliers could not increase their output in the immediate time-frame that Portal required, causing a major supply-chain problem for Portal

Norton Michaels was responsible for the procurement department at Portal. The buyers in his department worked hard to get suppliers to commit to the increased schedule. When his buyers were not able to get commitments from all their suppliers, he asked for a list of the suppliers who couldn't meet Portal's new schedule.

In an intimidating tone of voice, Norton called the suppliers on this list and gave them his minimum requirements. Not once did he reveal the truth about Portal's internal forecasting problem. He threatened each supplier with cancellation if they could not meet Portal's new schedule. These suppliers, forced with the prospect of losing a big customer like Portal, tried everything to improve their delivery schedule. Some resorted to decommitting quantities scheduled for their other customers in favor of giving these quantities to Portal. Other suppliers were simply not able to do enough to meet the steep increase in Portal requirements.

Time was running out for Portal. If Norton did not get enough materials from the suppliers, then Portal's production lines would be shut down. This would bring about layoffs and lost revenue, which would cause Portal to miss its quaterly earnings target.

Norton, was relentless in his attempt to get the materials needed to meet Portal's real schedule. He manipulated events to make it look like Portal's suppliers were responsible for missing its production target. He never admitted to any of the suppliers that Portal's forecasting software was the real cause of the problem. He never invited their assistance in resolving a problem that required the mutual effort of Portal and its suppliers. Norton conveyed arrogance and haughtiness and, in turn, invited his supplier's minimum cooperation. His misbehavior caused suppliers to do only what they had to do, and no more. The fact is, some of these suppliers could have done more to help Portal, had Norton conveyed less arrogance. But, Norton invited his supplier's minimum cooperation, because he did not convey an honest and sincere message.

Accordingly, Portal's suppliers did not frequently update Norton with new information because they were either too scared to contact him or they had decided to do as little as possible. This lack of new information caused Norton to become even more difficult as he badgered his suppliers. Norton's stress level was abysmal. He had trouble sleeping and he wasn't eating right. His life seemed consumed with overcoming Portal's scheduling problem. It never occurred to him that much of his stress *could have been averted if he conveyed a message that invited more mutual cooperation.*

Portal did not hit their desired production schedule. Norton and his buyers were held responsible for this catastrophe. When Norton was confronted about his department's inability to hit Portal's production target, he naturally blamed his suppliers. This debacle stained Norton's standing and career potential with Portal's management. He was eventually demoted to a postition where his actions would have less impact on Portal's business success.

Norton Michael's, in the preceding story, had learned to get things done by bulldozing through people. Clearly, he was more interested in results than in preserving business relationships. He rationalized to himself that his behavior was acceptable providing he reached the desired outcome, no matter who got hurt in the process. In the end, he learned that he was working against a very simple principle: *We Invite The Type Of Behavior That We Convey Towards Others.*

When people convey arrogance, they invite commotion. Anger, conveyed to another, invites retaliation. This common-sense principle, when applied, will enable a customer service professional to experience less stress because they'll never get stressed in the first place.

Conveying a bad attitude is like hurtling darts at the other person. This type of behavior has

Convey Darts

and you will ...

Invite Daggers

Convey Niceness

and you will ...

Invite Cooperation

consequences. After those darts have been thrown, be prepared for the daggers that will soon be aimed at you.

So, ask yourself; "Am I inviting more stress into my life by conveying a bad attitude?" If the answer is yes, then it's time make a choice to convey a more professional and courteous demeanor in all your business and personal interactions. This positive choice will not only benefit those you communicate with - it will also alleviate much of your own stress.

Never Give Up

As you go through life, you may find yourself at one of these learning stages regarding all the different activities that you engage in. You might start small, improve on a few activities, then move on to subsequent stages of learning and development to the final stage of unconscious competence.

However, there are other activities in which you never get passed the second stage of conscious incompetence. You either lose interest or give up out of frustration.

Improvement will not happen without your hard work. You must put something into it to get something out of it. Remember, each stressful experience we overcome prepares us to handle the stressful experiences that lie ahead.

Don't be like the man who stands in front of a fireplace and says, "Give me some heat, and then I'll give you some wood." It doesn't work that way. This man must first get busy gathering the wood before he can expect any heat from the fireplace.

Action is always necessary to get things done. So get busy! If you want to make positive changes in your life, you must unlock the gate of change in your mind and allow new knowledge to enter through it. I cannot open your gate of change. Only you can open your gate from the inside. You have the key.

Along with action, innovation is essential. When I consider innovation, I think about my parents. They were the most innovative people I ever knew. Allow me to give you a true example. When I was sixteen years old, I went to local roller skating rink with some friends. This was my first roller skating experience, so it took me about twenty or thirty minutes to get the hang of it. However, once I did I was having a load of fun. I started to get a little reckless and entered a turn at a high speed and lost my balance. I extended my right hand towards the floor to buffer my fall and this resulted in a broken wrist.

I went to the hospital where doctors wrapped my right arm in a cast from my right hand all the way to my shoulder. When I arrived home that Sunday afternoon, my mother had some initial concerns about my well-being. However, since I was OK outside of the broken wrist, she walked me into our dining room. She asked me to sit at the dining room table while she went to the other room to get a writing pad and pen.

Speaking in her calm, authoritative tone my mother reminded me that Monday was a school day. She told me that my condition would not keep me out of school. Then she handed me the writing pad and pen and she suggested that I learn to write with my left hand so I wouldn't fall behind in any of my school work.

I sat at the dining room table all Sunday afternoon and evening writing with my left hand until my left hand writing became fairly legible. In subsequent days, my left hand writing became as fast and legible as my right hand writing.

I take no credit for the solution my mother proposed. It was all her idea. However, I did benefit from learning about resilience and innovation.

I learned that obstacles are ever present. Obstacles have a way of setting us back temporarily, but if we have a resilient personality along with innovative thinking, nothing is impossible. This experience has blessed me with an attitude of perseverance and tenacity.

Chapter 3 Key Points

- Put new ideas and strategies into immediate action.
- Learning occurs in four stages of development that result in personal mastery.
- The four stages are, Unconscious Incompetence, Conscious Incompetence, Conscious Competence and Unconscious Competence.
- Improvement will not happen without your hard work.
- Each stressful experience you overcome prepares you to handle the stressful experiences that lie ahead.
- Practice innovative thinking as a strategy for overcoming obstacles.
- We invite the type of behavior that we convey towards others. If you convey niceness, you will invite cooperation.
- Never give up. A person with a resilient personality will overcome life's temporary setbacks.

Chapter 4

Rational Thinking

How do you feel when you're handling a call from an irate customer? Have you ever concentrated on the metabolic changes going on inside your body? Try doing this: Pay attention to the way you feel the next time you handle a call from an irate customer. Get in touch with your body and fully comprehend what's going on. I have done this; here's how I felt the last time I handled one of these calls.

The first change I noticed was a feeling of nausea in the pit of my stomach, followed by an increase in my rate of breathing. Then my voice changed. It became thin and raspy. These changes were followed by my survival instincts which urged me to get to safety as soon as possible. This means I felt pressured to do what was expedient, not what was correct. I just wanted to get away from whatever was threatening me with harm. I wanted the situation to end immediately. All this took about two or three seconds.

Then, by habit, I started thinking rationally. I matched what I knew about the metabolic changes caused by stress to what was really happening. My body's reaction to this adverse event was consistent with what I knew. No surprise there. No need to get upset or stressed out.

Then I consciously slowed down my rate of breathing, stood up and paced in my cubicle, took a sip of water and began the process of containing the situation.

The events I just described have become second nature to me. It has become a habit. I barely have to think about what to do.

After each event is successfully resolved, my mental database is updated with another positive outcome that will help me handle future ones. These successfully handled events have a cumulative affect on how I will think and behave later on.

The rational thinking techniques I use force me to focus on *what to do* rather than *what is happening.*

Focusing on what is happening will only result in negative feelings about the customer and his tirade. This irrational thinking leads to a downward spiral of inappropriate and defensive behavior which will result in a bad situation getting even worse.

Learning to change the way you think starts with gaining new knowledge. If this book changes the way you think about customer service, stress and irrational thinking, don't stop now. Continue your quest of new knowledge and keep adding to your database of experiences.

Grace Under Pressure

Michele Logan is the customer service manager for DDZ Inc., a computer manufactur - ing company. On a Friday afternoon, Michele received a voice mail from an irate customer. The customer, Sean Vorian, was a real nightmare. He bought a DDZ computer a few months ago and he has done nothing but complain ever since. Mr. Vorian was a chronic complainer with a history of using legal means to achieve his interpretation of satisfaction. Every time Michele attempted to resolve a problem with Mr. Vorian's computer, there wasn't a problem to solve. The computer worked just fine. Still, Michele worked along with Mr. Vorian during each new complaint.

This particular Friday, Mr. Vorian's computer was back at the factory again for evaluation. The computer arrived at DDZ on Thursday, the day before. Mr. Vorian's voice mail message demanded immediate return of his computer via overnight freight for Saturday delivery. Michele looked into the computer's diagnostic evaluation and while there wasn't any problem found, the responsible thing would include letting the computer run on a

---

diagnostic program for at least 48 hours. Michele knew that shortening the evaluation time, by returning the computer, would circumvent the diagnostic process and prevent her from learning about any potential problems.

Michele prepared to return Mr. Vorian's call. This was Michele's least favorite part of customer service management. She was prepared to disappoint Mr. Vorian with the truth. Michele learned that it never paid to fudge facts or satisfy a customer with a lie.

Before making the return call, Michele wrote a quick script to guide his thoughts during the telephone call. She had much experience in handling difficult calls. She spoke in a decisive and authoritative tone, without any gaps or hesitation due to word grasping. Michele finished her script, dialed Mr. Vorian's number and began the call in a polite, professional and authoritative manner. Michele described, in absolute terms, the nature of the diagnostic program and how it worked. Michele explained that it was in Mr. Vorian's best interest to be patient and wait until Monday.

Mr. Vorian's demand for an immediate return evaporated into a clear understanding of the situation and a request for immediate follow-up on Monday morning. Michele agreed to call Mr. Vorian on Monday morning with the diagnostic program's results.

The preceding story relates the importance of rational thinking, good preparation and positive expectancy. Complex problems, such as the one in the previous story, require a proactive and rational strategy. CSRs that use these skills experience less overall stress because they never get there in the first place.

Think Before You Work

By now, you have probably learned how important rational thinking is to improved stress handling. A stressed out CSR believes his biggest problem is the customer's behavior. This is wrong. This type of irrational thinking gets CSRs into more trouble than almost anything else. Rational thinking is a critical part of doing any job well.

Too often, CSRs answer an incoming call or make an outgoing call without thinking. Doing this is like driving blind. Sometimes CSRs go through all the involuntary motions of picking up an incoming call, using the proper greeting and expecting the call to handle itself automatically. While it's true that most calls are typical and require little or no effort, every now and then an unusual event occurs. Maybe it's an irate customer or maybe it's something else.

These unusual events catch an unaware CSR off guard, disrupt his lethargic state and instigate the stress mechanism. In this state of stunned surprise, a CSR is frequently forced to react, rather than respond. Not thinking will almost always result in reactive behavior.

In the beginning of this book, I wrote that we are big on *doing,* not *thinking.* Our focus is on getting things done. If CSRs were dealing with things, they

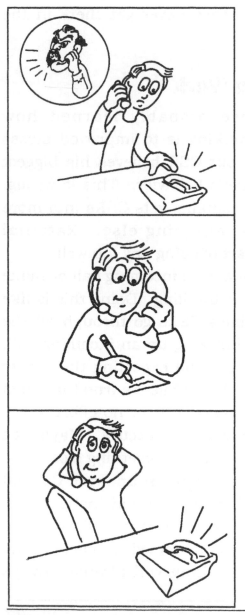

Before you pick up a new call, ask yourself - am I ready for anything?

If you are prepared and ready, you will think rationally and use the correct skills so you can overcome adversity.

This leads to a positive cycle of well handled events that prepare you for the next call.

TELE-STRESS

might be able to behave in a more lethargic manner. That's OK because things are predictable. However, people (our customers) are not predictable. One customer's minor inconvenience is another

If you answer the next call without thinking, you are not ready.

You will get caught off guard and be taken by surprise.

This leads to a negative cycle of mishandled events and poor performance.

customer's major catastrophe. Our customers are all different. Most of them are nice and a few are nasty, but we have to be ready for all of them.

Always make time to think, prepare a short script and have a liquid refreshment handy prior to making a difficult call.

Physical Activity

Physical conditioning is an important part of handling stress well. I have a great deal of personal experience in this area. Being physically active is one of the proactive things that adds to a positive outcome.

Structural engineers have learned that pre-stressing materials, such as steel I-beams, steel corrugated flooring, and steel cabling, before the assembly of buildings and bridges is beneficial. Pre-stressing materials, by applying the loads that will come when the structure is completed, will enable that structure to better withstand pressure when it is completed and in use.

Good physical conditioning can accomplish the same thing. A person who is in good physical condition will withstand the physical changes that arise during stressful events. In addition to proper exercise, good diet habits and a good mental attitude are also important.

When our whole body is healthy, both physically, mentally and spiritually, we stand a much better chance of withstanding stressful events.

Personally, I have been rewarded for the

exercising I have done during the last decade or so. While going through some of my toughest customer service challenges, some of which lasted days, it was my rigorous exercise program that helped.

Before starting an exercise program, consult with your physician. There are also numerous books, audio tapes and videos on this subject. Be a learning individual and keep building new knowledge.

In addition to exercise, make sure you get enough rest. If you have trouble getting to bed early, set your alarm for bedtime as a reminder.

Proper Diet

Another proactive strategy for handling stress, includes eating and drinking sensibly. When we get stressed, the sympathetic nervous system gears up to help with the fight or flight response by releasing two hormones, adrenaline and noradrenaline. These two hormones cause the heart rate to speed up, blood pressure to increase and the rate and depth of breathing to increase to help the lungs deliver more oxygen and nutrients to the working muscles. All these processes take energy (food) which is why it's important to eat a healthy, well-balanced diet.

Watch your sugar intake. Sugar may give you a quick energy boost, but the effect soon wears off leaving you more tired, irritable, anxious or depressed. In addition, an excess of simple sugars in the body tends to deplete the body's vitamin stores. Eat more fruits and vegetables to maintain a balanced diet.

An important note about stress is that too much caffeine will cause your adrenal gland to overproduce adrenaline. Therefore caffeine is a substance that can trigger the stress response. Caffeine speeds up your nervous system which results in hyper-alertness, often making you more susceptible to perceived stress. Caffeine is found in soft drinks, coffee, tea and chocolate.

Monitor your coffee intake. Personally, I give myself a two cup limit. This is just enough to jump start me in the morning, and it doesn't appear to deteriorate my ability to think clearly.

Alcohol is another substance we should use wisely. Although alcohol seems to have a calming effect on the body, it is actually a depressant. Alcohol numbs the body's systems and numbs our perception of things. Excessive alcohol consumption depletes important vitamins and minerals needed for the body's stress response.

Always eat a good breakfast. Your body needs energy after going about eight hours without food. Skipping breakfast can make you feel tired and it can cause headaches. Eating a good breakfast can give you energy for the day ahead and make you alert for the days challenges. Especially the dreaded morning calls.

On days when I eat inappropriately, I notice a deterioration in my ability to think clearly and contain adverse situations.

Self Talk

We talk to ourselves. This is called self talk. Customer service professionals have a nasty habit of engaging in very negative and condescending self talk. Sure, we all make mistakes - but this doesn't mean we have to beat ourselves over the head each time something goes wrong. Learn to change your negative self talk to positive self talk.

Most CSRs will replay bad experiences and mishandled events in their head, over and over again. Doing this lets them re-live that negative outcome, not just once, but numerous times. This is irrational thinking, and this habit will have a cumulative and negative affect on their performance and their personal behavior.

Rational thinking includes the ability to change this bad habit. Stop yourself from engaging in negative self talk. When you find yourself doing this, stop and change your negative self talk to positive self talk. Find ways to acknowledge and praise yourself for things you have done well. Focus on all your positive outcomes. For every one thing that goes wrong, there are probably 10 or 50 or 100 things that go right. Be realistic and count your blessings.

You will find that changing this habit is similar to going through the four stages of learning mentioned in the previous chapter. At first, it will seem uncomfortable and difficult. However, the more you apply yourself at thinking positive, the easier it will become.

Using positive affirmations is an easy way to change your negative self talk to positive self talk. Since our mind can only hold one thought at a time, force yourself to say something positive about yourself. Say something like: "I am an excellent call center professional," or "I solve problems effectively," or "I enjoy helping customers." Psychology experts propose that affirmations must be positive, present tense and personal. Do not use negative words or future tense in your affirmations. Here is an example of what you shouldn't say: "I will NOT get upset the NEXT time I handle a difficult call."

Using positive affirmations is a great way to divert your mind's attention away from negative thoughts so you can work more effectively. I have experimented with affirmations after a difficult day of handling irate customers. I use the affirmations in my car on the commute back home. These affirmations keep me from wasting energy on an activity that yields no benefit whatsoever.

The most important step in this process is the beginning. You must start. Make a personal commitment to start changing the way you think. Your thoughts are a reflection of who you are. Each person eventually becomes what they think about. People who view themselves in a negative context limit their chances of growth and success. If you change the way you think and start seeing yourself as positive and capable, then that is what you will eventually become. It all depends on how you think.

This Is Reality

Ben Tesche had to leave his desk for a moment and photocopy a technical sheet for a customer. In the corridor, he met Mary who reminded Ben about a report deadline that was due tomorrow. Ben remembered that he needed some research documentation before he could complete that report. "I might as well get it now," he thought. So instead of going to the copy machine he headed for the documen - tation department.

On the way, Ben heard himself being paged on the PA system. He answered the page and learned that his boss asked him to return to the call center for an urgent call. Wanting to get the documentation first, he asked to have the call switched to the phone he was at. He took the call, wrote notes on the back of the page he was going to photocopy and promised to call the customer in a few minutes with a response.

When he got to the documentation department, Ben couldn't find what he needed. The clerk told Ben that someone from engineering was using the documentation and would return it sometime this afternoon. Ben said he would return later and remembering why he left his desk in the first place, went to photocopy that technical sheet. Unfortunately, the photocopy machine was broken and wouldn't be fixed for a few hours.

Ben went back to his desk, without his photocopy or the documentation and returned his customer's call.

Sound familiar? How many times do you find yourself running around, expending energy and not really getting anything done. Unfortunately, this is reality. Most call center professionals do much multitasking. Juggling these multiple tasks can add stress to our lives.

While situations such as these cause much frustration and anxiety, don't allow yourself to get too stressed out over it. It's best to make a conscious choice to think rationally and move on to the next task. Rational thinking includes knowing when to invest time and energy improving a situation and when to accept situations as they are. I urge you not to waste energy becoming stressed out over things you can't change.

Today's business environment, with multitasking and modern complexity, is busy and chaotic. This will not be changing soon, so we ought to change the way we perceive these events.

Ten Strategies for Staying Courteous Under Stress

1. The Rudiments - Your mother probably conducted your first and most important communication skills seminar. She taught you to say "Please" and "Thank You." These rudiments still work wonders. When stress renders us ineffective, the rudiments are often the first thing to go. In addition, always tell the truth. Do nothing you have to lie about later. Telling lies will only add to your stress when the truth eventually surfaces.

2. Identify Stress Signals - Learn to recognize when you are under stress. It is that hyperactive feeling you have after closing a call with an upset customer. You know, that tightening in the chest or feeling like you are out of breath or like you just have to get up and take a walk. The stress mechanism gives you an immediate burst of energy, so you'll want to do something physical, it's only natural.

3. Proper Breathing - Force yourself to breathe slowly. This is important. Doing this consciously forces you to think rationally about what just happened. You had an unpleasant event that forced your metabolism to increase. Breathing slowly will help offset the sudden change in your metabolic rate and allow you to remain courteous. Learning to respond to adverse events by breathing slowly will eventually become a habit. This habit will enable you to think rationally rather than give in to your body's fight or flight response.

4. Rational Thinking - Thinking rationally about what caused you to get stressed is the first step towards keeping things under control. You are probably not being threatened with bodily harm, yet the stress mechanism makes you feel as though you are. Challenging your own thought process can help you overcome the affects of stress in customer service. Our society is very big on not thinking. We are very big on doing. Action is seen as the first and foremost activity. Thinking is secondary. We feel it's more important to get the next call rather than to take a few moments to do a ten second readjustment. Since irrational thinking produces stress, doesn't it make sense that rational thinking will help manage it? Investing a few seconds in rational thinking will pay off with improved performance and increased customer satisfaction.

5. Change your Environment - Take a break when you really need one. When stress renders you ineffective, change your scenery. Find a window, and look for something peaceful like a tree, a flock of birds or a garden. These short, five minute, readjustments will revitalize you and enable you to regroup so you can continue working, more effectively.

6. Learn to Forget - Don't replay bad experiences in your head, over and over again. Doing this reduces your self-esteem and drains much needed energy on an activity that yields absolutely no benefit. In addition, it distracts you from the customer you are presently working with and it may cause you to appear discourteous. Substitute rational thinking for this self-destructive behavior. Your mental attitude will affect your behavior, so maintain a positive outlook on your job.

7. Wet Your Whistle - Keep a liquid refreshment such as fruit juice or water on your desk. I know this sounds silly, but you will need it. Here's why. One of the results of the stress mechanism is a change in your voice. The voice becomes hoarse or raspy when you become stressed. Since your voice is your primary communication tool, you should ensure it is always operating at peak performance. Doing this will enable you to maintain a calm, consistent and courteous tone of voice.

8. Accentuate the Positive - You will appear more courteous when focusing on a positive outcome instead of a negative one. For example, telling a customer, "This isn't as bad as your last problem" instead of, "This is much better than your last situation," focuses on the down side of a situation. You should direct the customer's attention towards an optimistic future by reinforcing the positive potential of a situation, not the negative possibilities.

9. Don't Blame Customers, Assist Them - You should be careful not to place blame, even when the customer does something wrong. If after a customer describes a problem, and you qualify that the problem is a result of something the customer did, and you say, "You're wrong! It doesn't work that way," instead of "I may be wrong, but I believe it operates a little differently. Please allow me to help," the problem can be exacerbated. Words are weapons. The words you use, during these events, will make you appear either courteous or antagonistic. Substituting the word "I" for "You" takes the emphasis (and blame) off the customer and neutralizes the event.

10. Do Not Retaliate - Use restraint not retaliation. Some customer's only concern is satisfaction, which means getting aggressive to get what they want. When you are working with rude or abusive customers, and you feel your stress mechanism activating, use restraint instead of retaliation. An angry customer's behavior is unpleasant, but if you feel the temptation to retaliate, don't. This will only escalate the situation, not contain it. Behave like a professional. Speak in a calm consistent tone and help the customer conform to your professional behavior.

If you retaliate, one of two things will happen. One, the customer may counterattack more severely and the situation will escalate out of control. This will make your stress even worse as the customer becomes more difficult to work with. And two, you will damage the customer's self-esteem and force him to submit for the short term, but your company will lose the customer for the long term when he purchases some place else the next time.

Chapter 4 Key Points

- Think before you work. Invest a few seconds in preparation.
- Live a moderate lifestyle and get enough rest.
- Good stress management includes eating a healthy, well-balanced diet.
- Too much sugar will render you tired, anxious, irritable or depressed.
- Caffeine speeds up your nervous system which results in hyper-alertness.
- Stop negative self talk. Become disciplined at substituting positive self talk until it becomes a habit.
- Use positive affirmations to improve your conscious thought and overall mental outlook.
- Some situations are not worth getting stressed out over. Learn to accept reality and move on rather than languish in anxiety and frustration.

Chapter 5

The Stress Survey

In preparing to conduct a survey on call center stress, I used a short and effective questionnaire. Previous surveys I conducted revealed that there is a relationship between the size of a questionnaire and the return rate. Short questionnaires yield a higher return rate than long questionnaires.

Respondents were guaranteed their anonymity. Companies were selected from the following industries: Health Care, Power Utilities, Food Service, Car Rental and Home Furnishings. The aggregate was comprised of 158 respondents.

Each respondent prioritized their top three causes of stress and their top three types of post-stress behavior. Having respondents list their top three choices would allow them to think about numerous issues, then compare and rank these issues in their order of priority. In an effort to capture primary causes and effects, I only reported on first choices.

This report is divided into two sections: Section one is a report of the aggregate responses to survey questions one through four. Section two is a more detailed analysis.

The second section uses question three as a qualifier and reports on questions one, two and four. How respondents answered question three put them into one of two groups: (1) those who believed they needed help handling stress and (2) those who believed that stress wasn't a concern. Section two reveals the subjective nature of stress. Namely, that the causes of stress, post-stress behavior and daily stressful events are contingent on whether or not a person sees himself as "needing help."

These results reinforced the concept that no two people react to stressful events similarly. One person's mild inconvenience is another person's major catastrophe.

Confidential Call Center Stress Survey
This is not a test, so there are no wrong answers.
Please think carefully and answer honestly.

Thank you! Stephen Coscia
P.O. Box 786
Havertown PA 19083

1 **Top 3 causes of stress:** Select your 3 most stressful events, then rank all three. Write in boxes.
(1 = 1st place, 2 = 2nd place, 3 = 3rd place)

☐ Calls from chronic complainers

☐ Calls from irate customers

☐ Calls from intellectually challenged customers

☐ Too many interruptions at work

☐ Personal or family situations that interfere with work

☐ Other: _____

☐ Other: _____

☐ Other: _____

2 **After a stressful event:** Select your top 3 answers, then rank all three. Write in boxes.
(1 = 1st place, 2 = 2nd place, 3 = 3rd place)

☐ I feel angry ☐ I feel depressed ☐ I need a drink ☐ Other: _____

☐ I need a long walk ☐ I need to talk with someone who cares ☐ Other: _____ ☐ Other: _____

3 I want to learn more about my body's metabolic changes during stress. (Yes / No)
Why? _____

4 I experience approximately _____ stressful events daily. (How many each day?)

The survey sheet was short and simple.

Section 1

Causes Of Stress

There is a correlation between causes of stress and calls from irate customers. Based on the survey, irate customers cause more stress in call centers than anything else. Almost half (47%) of all respondents listed calls from irate customers as their number one cause of stress. Other first choice responses were divided among seventeen other categories. After irate customers, the next three most common responses include interruptions (15%), dumb customers (7%) and personal situations (6%). The four categories listed above make up 76% of the aggregate.

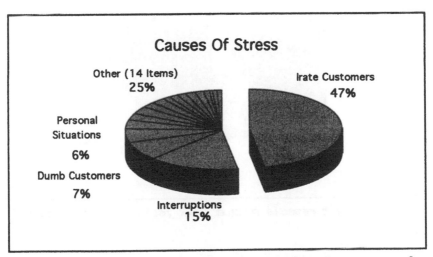

Irate customers are, by far, the number one cause of stress among customer service professionals.

Post-Stress Behavior

The answers to this question about post-stress behavior reveal how a CSR feels after a stressful event. Since we already know that irate customers are by far the number one cause of stress, it stands to reason that many CSRs selected anger as their first choice for post-stress behavior. Anger and frustration made up 28% of the responses to this question, followed by talking with someone who cares (21%), taking a long walk (15%) and feeling depressed (8%). These four categories make up 72% of the aggregate.

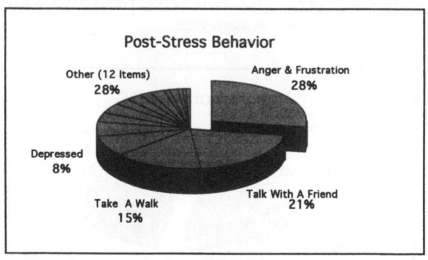

This chart reveals how a CSR feels after a stressful event.

Daily Stressful Events

Ninety eight percent of the respondents experienced at least one stressful event daily. When asked about the quantity of daily stressful events, 25% of the respondents said one event, 14% said two events, 17% said three events, 9% said four events and 15% said five events. These five categories make up 82% of the responses. The remaining answers ranged from six through fifty. The highest numbers appear to be exaggerated.

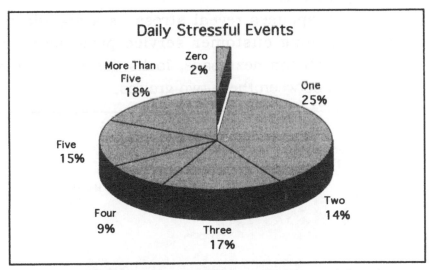

Almost all respondents (98%) experience at least one stressful event daily.

Learn More About Stress?

Most of the respondents (63%) said they wanted to learn more about stress. Among the 63% who

Yes	63%
No	37%

answered "yes", 83% also included a reason why. These included health reasons (39%), work and productivity reasons (33%) and general curiosity (11%). Seventy two percent of the respondents who answered "yes" are concerned about their health or their productivity.

These responses reveal stress as a significant concern among customer service professionals. Section 2, on the next page, looks deeper into the differences between these two groups.

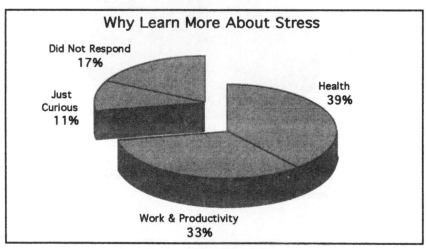

72% of the respondents who answered "yes" are concerned about their health or their productivity.

Section 2

Sensitive and Pragmatic Respondents

Section two reveals the differences between the respondents who answered "yes" or "no" to whether or not they wanted to learn more about stress. Respondents who answered "yes" see themselves as needing help. I named this group *sensitive*. These respondents have a greater sensitivity to stressful events. The data reveals that their reaction to stress is more severe than their counterparts'.

Respondents who answered "no" are referred to as *pragmatic*. This group appears to need less help. The data reveals that pragmatic respondents are less shaken up by irate customers, their post-stress behavior is more practical and they experience fewer daily stressful events.

This data reinforces the subjective and relative nature of stress on individuals. The cause of stress, post-stress behavior and quantity of daily stressful events are contingent on whether or not a person sees himself as "needing help."

Causes Of Stress

The differences between causes of stress for sensitive and pragmatic respondents were impressive. The sensitive respondents were almost twice as likely to indicate irate customers as their number one cause of stress. More than half (55%) of the sensitive respondents compared to only about one third (35%) of the pragmatic respondents indicated that irate customers were their number one cause of stress. The sensitive respondents evidently view irate customers as a major cause of stress. The pragmatic respondents might have a higher tolerance for these types of calls, or they may have more resilient personalities. They're obviously not as easily shaken up and disturbed by irate customers.

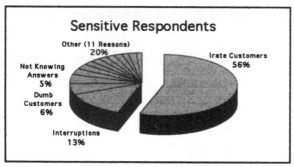

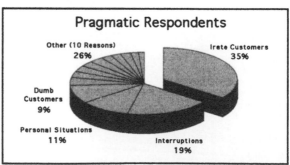

Post-Stress Behavior

There are significant differences between the post-stress behavior of sensitive and pragmatic respondents. The sensitive respondents were more likely to get angry after a stressful event. Almost one third (32%) of the sensitive respondents, compared to only about one fifth (20%) of the pragmatic respondents, indicated they became angry after a stressful event. In addition, the pragmatics cited more overall constructive post-stress behavior. The pragmatics were more likely to talk to a caring person or take a walk rather than become angry.

Some of the destructive post-stress behavior exhibited by the sensitives included a higher propensity to become depressed or smoke a cigarette.

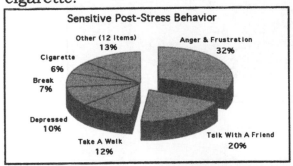

Sensitive Post-Stress Behavior

Other (12 Items) 13%
Anger & Frustration 32%
Cigarette 6%
Break 7%
Depressed 10%
Take A Walk 12%
Talk With A Friend 20%

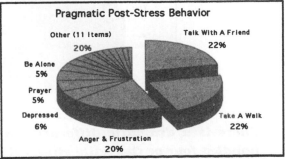

Pragmatic Post-Stress Behavior

Other (11 Items) 20%
Talk With A Friend 22%
Be Alone 5%
Prayer 5%
Depressed 6%
Take A Walk 22%
Anger & Frustration 20%

Daily Stressful Events

The differences between daily stressful events for sensitive and pragmatic respondents are also relevant. Overall, the sensitive respondents experienced a greater number of daily stressful events than their pragmatic counterparts. Clearly 31% of the sensitives indicated they experienced either four or five daily stressful events compared to only 14% of the pragmatics. The pragmatics were 45% more likely to experience only one or zero daily stressful events. Not one sensitive respondent indicated zero daily stressful events.

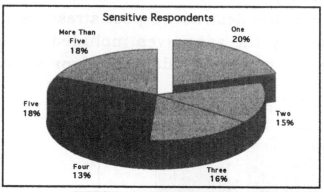

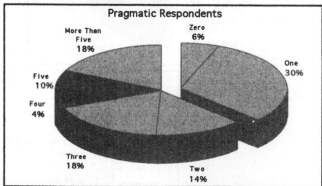

There is a dramatic difference in respondents who handled four or five daily stressful events.

Survey Summary

I didn't plan on the survey results specific to sensitives and pragmatics. This happened by accident. While reviewing and documenting the reasons for why respondents did or didn't want to learn more about stress, I noticed a common trend: respondents who were interested in learning more about stress were also most likely to cite irate customers as a cause of stress and, on average, they experienced a greater number of daily stressful events.

This revelation inspired me to sort all the survey responses by the "yes" or "no" answers to question 3. This may be the first statistical evidence on the subjective nature of stress in the call center environment. This data might be best utilized by human resource professionals. Screening for candidates with the most resilient personalities will result in less turnover, less burnout and greater return on training investment.

The survey results are valuable. Having this data quantified will make it more meaningful to call center managers and trainers. One fact is clear: Tele-Stress is a real problem that demands attention and real solutions. The vitality and productivity of customer service professionals stands in the balance.

Screening For The Best CSRs

My call center stress survey revealed two distinct types of CSRs: sensitive and pragmatic. This revelation should be of interest to call center managers who are looking to hire the best CSRs. Call center managers should remember that calls from irate customers and numerous internal interruptions disrupt and shake up sensitive CSRs. In addition, sensitive CSRs see themselves as "needing help." They have less resilient personalities and they have a propensity to become less productive after handling a stressful event.

Pragmatic CSRs are more stable and productive. They experience fewer daily stressful events and irate customers and internal interruptions don't disrupt them. Pragmatic CSRs have more resilient personalities and they see themselves as capable and autonomous workers.

Based on my survey, there are more sensitives than pragmatics. The ratio is about two to one. Screening candidates to find the 33% of pragmatic CSRs is a worthwhile investment for call center managers looking to recruit and hire the best.

This factor is so critical because of the cumulative affect that unproductive CSRs have on a call center. For example, if you manage a twenty seat call center and one quarter of your CSRs are operating at 60% efficiency, then you are losing the work output of two CSRs daily. In addition, sensitive CSRs are higher maintenance employees because they have a higher call escalation rate.

Screening and hiring good CSRs is a growing concern in our industry. Call center managers must only let the best people into their organization. Doing this is recommended because it relieves much work and frustration later on. I have heard many horror stories from call center managers who were on a mission to "fill seats" rather than screen carefully. Careful screening will help avoid obstacles such as productivity problems, inappropriate attitudes and training concerns later on.

I suggest an interviewing method that includes a combination of testing, listening and watching each candidate. I rate applicants in four areas: product knowledge, retail sales experience, communication skills and attitude. During an interview, I ask questions that give me a past performance picture which is the best indicator of future performance. Here's how I interview using my four rating factors.

Product Knowledge - I ask ten close ended questions, specific to my company's product or industry that have only one correct answer. This way the candidate must answer with (1) the correct answer, (2) the incorrect answer or (3) he'll say, "I don't know." I note whether the candidate enjoyed this exercise. The best CSRs don't just solve problems; they like doing it.

Retail Sales Experience - I look for candidates that have had some (even a little) experience working with the general public, preferably in a retail sales environment. This experience tells me that the candidate has a clue about different types of human behavior. The best CSRs understand that one person's minor inconvenience is another person's major catastrophe.

Communication Skills - I listen for subtleties in vocal tonality. I listen for too many uhs, slurred speech, fast talkers, slow talkers or other bad vocal habits. I role play with candidates by asking, "If I were a customer calling to place an order, and the item I wanted was out of stock, how would you respond?" The best CSRs use euphemisms and they turn negative situations into positive ones.

Attitude - I hire CSRs who REALLY want to work at my company. Candidates must enjoy working with customers and solving problems. The reward has to be more than just a paycheck. There must be a desire to overcome obstacles and persevere out of a need to help people. If working here is just another job, then the candidate might quit when the going gets tough. The best CSRs understand that persevering through the hard times instigate growth and development.

After interviewing candidates, I first rate them on a scale of one to ten (see chart). Then I distribute the weight of the scores as follows: product knowledge (15%), retail sales experience (25%), communication skills (30%) and attitude (30%). Weighted heaviest are communication skills and attitude because these traits are hard to change. It took years of behavioral conditioning to acquire these traits and this behavior will not change in few weeks or months. Retail sales experience deserves a factor of moderate weight as it is almost as important as the former. Product knowledge has the least weight. This learning is easiest so long as a candidate knows the fundamentals.

This process will help you make objective decisions while interviewing numerous candidate for a single position. These decisions are critical

Scores and Weight Distribution

	Product Knowledge	Retail Sales	Communication	Attitude	Total	
John Morgan	9	7	9	6	30	Candidates score from one to ten
Linda King	10	9	8	8	35	
Cecilia Wright	8	6	7	5	26	
Joel Abbott	9	8	10	6	33	
Weight Distribution	15%	25%	30%	30%		
John Morgan	1.2	1.8	2.7	1.8	7.45	Candidates final score after applying weight
Linda King	1.5	2.3	2.4	2.4	8.55	
Cecilia Wright	1.5	1.3	2.1	1.8	6.65	
Joel Abbott	0.9	2.0	3.0	2.7	8.60	

In the top matrix of the above example, Linda King had the highest score. However, after weight distribution, based on interviewing factors, she isn't necessarily the best candidate for the job.

because human assets are your most precious resource. For a new hire, the first few days are the most memorable. This is when a new hire will create a mental image of his new employer's permissiveness, disciplines, fairness, strengths and weaknesses. Once a CSR creates this image, it's difficult to change. Most bad habits start here. While you may not have the time to do all the training and development, delegate this important task to someone who can make a positive impression on new hires.

Call center managers have the awesome responsibility of navigating their organization in a challenging and sometimes hostile environment. This industry is challenging due to the dynamic nature of call centers. Call centers become hostile when irate callers pursue an adversarial relationship with CSRs. My experience in working with numerous call centers has revealed a trend among call center mangers who manage well. The best managers focus on two key long-term strategies: (1) leadership and (2) training.

Leadership

Resolving complex call center problems is not easy. Things are often uncertain and ambiguous. This condition makes some call center managers want to retreat rather than lead. Don't let this happen to you. If CSRs look around for leadership and don't see it, they'll start doing what's most comfortable for them. Unfortunately, this might not be what's best for the company.

The best call center managers make time to think. Throughout history, the greatest leaders went off alone and separated themselves from the crowds for a while. Accordingly, they were often the ones whom the crowds most wanted to follow. A manager must have some solitude so she can create a plan for how to overcome today's obstacles and create vision of where her call center should be next week, next month or next year. A leader is a person who plans her work, then works her plan.

Many call center managers don't lead because they don't know where they're going. A leader moves toward a better destination than those around her. Then, she shares a vision or message about this better destination. She communicates so well that others get in line, behind the leader, to get to that better place. Leaders with a well-communicated vision attract followers voluntarily and they achieve their goals.

Training

Call center managers must continually train in two areas: (1) verbal communication and (2) stress management. Why these two areas? This is because of the relationship between communication and stress. For example, CSRs who communicate poorly will upset and disappoint customers more than CSRs who communicate well. As a result, upset customers tend to yell and complain which results in more CSR stress. When a CSR becomes stressed, his communication skills deteriorate even more. This downward spiral will eventually lead to burnout.

Call center managers must continually invest in training programs that focus on skills such as vocal imaging, word usage, listening, problem solving, rational thinking and stress management. When CSRs have been trained to balance both their verbal communication and stress, they will satisfy more customers, increase their own job satisfaction and stay more productive.

Too many call center managers don't invest in training because they claim to not have the time or resources. This is a mistake. When it comes to training, it's best to believe that the cost of training is less than the cost of ignorance.

Chapter 5 Key Points

- Irate callers are the number one cause of stress among CSRs.
- The most common manifestation of post-stress behavior is anger and frustration.
- The best interviewing method includes a combination of testing, listening and watching each candidate.
- The ratio between sensitive and pragmatic CSRs is about two to one. Call center managers should strive to recruit the 33% of pragmatic CSRs.
- Call center managers must embrace a strategy of good leadership and ongoing training.

Chapter 6

Conclusion

My survey reveals the indisputable connection between calls from irate customers, and CSR anger. This connection is Tele-Stress. Much of the anger experienced by CSRs can be suspended using the techniques found in this book. External sources provoke and instigate anger in our lives. Since we can't change the external sources, it makes sense to work on the way we perceive these events.

I know how ugly and difficult some customers can be. When I conduct workshops, I hear horror stories from CSRs, about customer behavior. Yet, I am always amazed at the resilience of the CSRs who relate these tales. Customer service professionals are some of the most people-oriented beings on the planet.

As discussed in the introduction of this book, some of the impetus for customer's behavior might be attributed to our changing world. This will not stop. The world is going to continue changing and advancing at a faster pace. Looking back over the last 40 years, things have changed dramatically. In the 1950s, 73% of U.S. employees worked in production or manufacturing. Now less than 15% do. Today, more than half of U.S. employees work in the service sector. As our business economy is centered more and more on knowledge, as a key product, there will be frustration as some people simply won't be able to keep up.

I do not believe that CSRs should have an adversarial relationship with customers. That's not what this book is about. My definition of customer service is: "The delivery of information or services to customers by a communicator who is an expert in his field." Unfortunately, some events turn sour and this book is about how to deal with the stress of those events.

Personal Notes

If you have had an opportunity to put my strategies into practice, and received positive results - good for you. However, don't stop there. Take what you know and expand upon it. Your local library or book store has a wealth of information on stress relief, customer service, communication and business. I have found it very beneficial to listen to audio books while driving to and from work. Whatever you do - keep learning.

Throughout life, you will encounter problems that will seem unsurmountable. These problems might be personal or work related. Dwelling on problems for hours, days or weeks is itself very stressful. I suggest a common-sense strategy for gaining objective insight into these problems. Too often we are too close to our own problems. We're so wrapped up and tangled in the problem that we can't see an objective solution.

I suggest a strategy that will allow you to take a step back and view the problem from an objective slant. The Japanese have an expression that means "A distanced view of things up close." We need to employ this same technique to our lives so we can get a proper and objective view of our life.

One of the best ways I have learned to get this distanced view is to write the problem down on paper. Write a one sentence synopsis of the problem. Once you can define and name the problem clearly and succinctly, you've taken an important first step. Being able to see exactly what the problem is will prevent you from drifting and wasting resources on what the problem isn't. Doing this is consistent with the problem solving strategy of process of elimination.

Next, on the same piece of paper, make a list of all the possible solutions. No matter how trivial or far fetched they might be. Then, start to scrutinize your list. Soon, you will start to see solutions taking shape. This approach seems so rudimentary, but it's still so effective. This process of "naming a problem," then listing all possible solutions is another way of getting yourself to think. Again, we're big on doing - not thinking. So, when problems arise, stop what you're doing and start thinking.

I have also found it useful to keep a journal. I have been writing in a journal since 1975, and it has become a tremendous resource. There are few things more exciting that reading your own hand writing, from ten or twenty years ago about dreams, goals and aspirations, that you have since realized and accomplished. A journal is also a wonderful way to vent your feelings and frustrations. Doing this makes it easier to let go of problems.

I express my sincere thanks for allowing me to share some ideas and strategies with you on Tele-Stress. I hope this information is meaningful and that it finds a useful place in your business and personal life.

I invite your questions and comments about any this book's content. You may e-mail me via my website at www.coscia.com or phone me at 610-658-4417.

Index

A

B

C

H

Handling Stress Well 43

Hazardous Situations 24

Health 89

Health Care 87

Heart 22

Heart Rate 43

Heartbeat 22

Heightened Senses 22

Hidden Resources 61

Home Furnishings 87

Homeostasis 23, 24, 27, 45

Hostility 25

Hypothalamus 22

I

Identifying Problems 38

Ignorance 20

Inappropriate Behavior 36

Implementing Strategies 14

Improved Stress Handling 42

Inappropriate Behavior 32

Insulin 22

Interrupt Customers 38

Interruptions 40, 89, 90

Involuntary Motions 69

Involuntary Organs 22

Irate Customer 33

Irate Customers 28, 36. 37, 38, 42, 43, 45, 46, 86, 87, 90

Irrational Thinking 28, 32, 44, 69, 76, 78

K

Kidneys 22

Knowledge 19, 20, 21, 29, 57, 75

L

Lethargic Behavior 72

Learn More About Stress 89

Learning 57

Less Patient 14

Let Customers Vent 37, 38

Liquid Refreshment 79

Listening Skills 40

Liver 22

Long Walk 87

M

N

O

P

U

Unconscious
  Competence 58, 59,
  66
Unconscious
  Incompetence 58, 59
Unpredictable Behavior
  72
Unprofessional
  Behavior 36
Unskilled CSR 57
Upset Customers 39

V

Vengeance 34
Venting Anger 35
Venting Peak 37
Venting Process 36, 37,
  48
Verbose Customers 40
Vocal Changes 15, 25,
  28
Vocal Chords 45
Vocal Minority 13
Voice 44
Voice Gets Louder 37

W

White Blood Cells 22
Withstand Pressure 72

Host a Tele-Stress Workshop

Stephen Coscia conducts a limited number of Tele-Stress workshops each year. If your company is interested in hosting such an event, please contact Mr. Coscia and ask for a Tele-Stress workshop brochure.

Stephen Coscia
P.O. Box 786, Havertown PA 19083
610-658-4417
coscia@worldnet.att.net
www.coscia.com